MAGNESIUM

Magnesium is the lightest of the commonly used engineering metals, being 35% lighter than aluminium. Whilst world production of magnesium is at present only one fiftieth the weight of aluminium produced, the unlimited amounts of magnesium compounds available to mankind will ensure for the metal in the long term a progressively larger share of the light metal market.

This monograph reviews the properties of magnesium, traces the development of the principal alloys and outlines the main forming procedures for the metal. The chemical and engineering application are reviewed and examples given where the low density of magnesium alloys make them ideal constructional materials. Magnesium's role in nuclear power production is also examined.

Other titles in this series

ME/1 TRIBOLOGY
E D Hondros

ME/2 ELECTRON BEAM WELDING
M J Fletcher

ME/3 VACUUM BRAZING
M J Fletcher

ME/4 STRESS CORROSION FAILURE
P Greenfield

ME/5 GLOW DISCHARGE MATERIAL PROCESSING
R A Dugdale

ME/6 DIAMOND GRINDING
J Burls

ME/7 ENGINEERING APPLICATIONS OF BERYLLIUM
P Greenfield

ME/8 WHISKERS
C C Evans

ME/9 CREEP OF METALS AT HIGH TEMPERATURES
P Greenfield

ME/10 CAVITATION
I S Pearsall

Other Mechanical Engineering titles available

TL/ME/1 HYDROSTATIC EXTRUSION
J M Alexander and B Lengyel

TL/ME/2 THE HEAT PIPE
D Chisholm

All published by Mills & Boon Limited

M & B Monograph ME/11

General Editor: J Gordon Cook, PhD, FRIC

Magnesium

P Greenfield, B Sc, Ph D

Mills & Boon Limited

London

First published in Great Britain 1972
by Mills & Boon Limited, 17–19 Foley Street,
London, W1A 1DR.

ISBN 0.263.05057.2

Made and printed in Great Britain
by Butler & Tanner Limited, Frome and London

CONTENTS

1. Introduction

Magnesium metal was first produced by Humphry Davy in 1808, but it remained little more than a curiosity until the early part of the present century when the advent of heavier-than-air flying machines stimulated interest in all light-weight constructional materials. Magnesium alloys, which had high strength allied to a low density (about 1·74 g/cm^3) had an obvious attraction as materials for aircraft construction, and production of the metal increased.

During both World Wars, the explosive expansion of the aircraft industry, and the demand for flares, tracer bullets and incendiary devices, brought a rapid development in magnesium technology. In 1939, the annual world production of magnesium was 32,000 tons, most of the metal being produced by a war-conscious Germany.(1) By 1943, the figure had risen to 228,000 tons—a sevenfold increase in four years. When peace came, the demand for magnesium fell away with the cutback in military aircraft production. There was a huge stockpile of magnesium metal, and the amount of magnesium needed for peacetime applications was comparatively small. In 1946, world production of magnesium was only 10,000 tons a year.

In the post-war years, the demand for magnesium began to revive; by 1955 it had reached more than 100,000 tons, most of the metal being produced in the U.S.A. and U.S.S.R. In 1971, production was at the rate of more than 200,000 tons per annum. Despite the post-war growth of the magnesium industry, however, the weight of metal produced annually is equivalent to only one-fiftieth of the weight of aluminium produced annually. There is ample scope for the further development of magnesium as a peacetime engineering material, and there seems little doubt that production of the metal must continue to increase in the future.

The expansion of the primary metal-producing industry since 1956 is shown in Fig. 1, together with a regional breakdown of the total production.[2] This steady growth reflects the increasing use of magnesium for non-military applications.

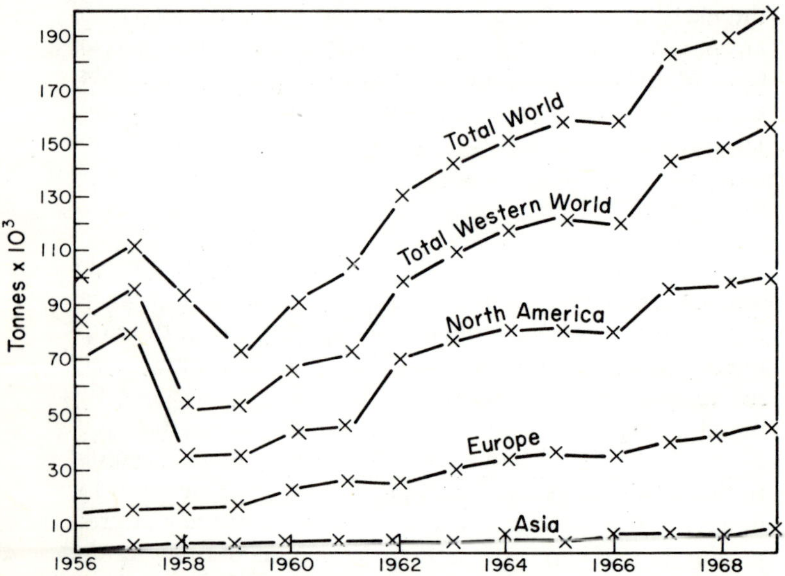

Fig. 1. Consumption of magnesium, 1956–1969.[2]

Magnesium compounds are found in great abundance, both as solid deposits and in solution in the surface waters of the earth. The most common ores are the carbonates, magnesite ($MgCO_3$) and dolomite ($MgCO_3.CaCO_3$). These minerals represent highly concentrated sources of magnesium, but the most abundant source is the magnesium chloride dissolved in seawater, which is equivalent to a magnesium content of 0·13%. This is a very low concentration, but the magnesium in the ocean is available in virtually unlimited quantity (one cubic mile of seawater contains some 5·25 million tons of magnesium); the magnesium chloride concentration remains almost constant, and permits of the use of highly standardised extraction procedures.

Magnesium is extracted using both electrolytic and thermal reduction processes, the former being commercially the more important. Electrolytic reduction processes in fact account for 80–90% of the current total output. These processes, employing fused anhydrous magnesium chloride as electrolyte, hold definite economic advantages where cheap electric power is readily available, e.g. from natural gas or hydro-electric sources. The heavy dependence of the economics of magnesium production on the cost of electricity has led to the demise of primary metal production in the U.K., where production was extensive during the 1939–1945 war.

The anhydrous magnesium chloride used as electrolyte is derived largely from magnesium oxide, and although chlorine is produced during electrolysis, the process has been a net chlorine user. Recently, however, the problems of dehydrating magnesium chloride to yield a satisfactory cell feed have been overcome and it is now possible to utilise the electrolytic reduction process to yield both magnesium and chlorine as co-products. Theoretically, 2·9 lb of chlorine are produced with every pound of magnesium. The ratio achieved in practice, which is dependent on the recovery efficiencies, varies between 2·9 and 1·8.[2] The value of the chloride is affected by the location of the plant and the presence of chlorine-using industries in the area, but it can reduce the price of magnesium ingot by as much as 20%. Less than 5% of the magnesium produced at present (1971) is yielding chlorine as a co-product, but as new plant is built or old plant modified, the value of the chlorine will make magnesium increasingly competitive with rival materials.

Magnesium is still produced in some parts of the world by direct thermal reduction of oxides with carbon or silicon under vacuum, but these methods are competitive with electrolytic extraction only under rare conditions of great natural site advantage.

2. Physical Properties

Magnesium (atomic number 12; atomic weight 24·32) is the lightest of all the engineering metals, having a density of only 1·74 g/cm^3. It is 35% lighter than aluminium and over four times lighter than iron and steel. Its melting point is 650°C and it freezes into a close-packed hexagonal crystal structure. Its absorption cross-section to thermal neutrons is only 0·069 barns and it absorbs neutrons at a rate which is almost 90 times less than an equal volume of steel. Magnesium can thus be used in reactor cores with only minor effects on the efficiency of the fission process. Magnesium alloys have consequently been used widely in commercial, natural uranium fuelled, gas cooled reactors as fuel element cans and supports. It is nevertheless the low density of magnesium which provides its main attraction, and cost savings arising from the use of light alloys in moving structures such as aircraft, motor cars, trucks, conveyors, etc., have brought about much of the increasing demand for the metal.

3. Alloys

Magnesium, which is readily obtainable commercially at purities of better than 99·8%, is rarely used for engineering applications in the unalloyed form, largely because of its low ultimate tensile strength. This is around 11 kg/mm^2 for sand castings and 20 kg/mm^2 for wrought products, with ductilities usually in the range 4–8%. It has been found that small alloying additions can improve these properties significantly. Whilst many alloying elements have been examined, only a small number are of outstanding interest commercially. These include aluminium, zinc, manganese, zirconium, thorium, the rare earths and beryllium, which are added singly or in various combinations in quantities small enough to ensure that there is no serious impairment of the light weight characteristics of the material.

The earliest commercial alloying additions were aluminium and zinc. These act as solid solution hardeners and, in the case of aluminium, in some measure as a grain refiner; intermetallic compounds which form also have some effect in promoting strength. Zinc is much less effective than aluminium in limiting grain size, and magnesium–zinc castings also showed marked tendencies towards a high degree of microporosity. It was not until the subsequent discovery of the exceptional grain refining characteristics of zirconium that the full potential of zinc additions could be realised in magnesium–zinc–zirconium ternary alloys. Nevertheless, small zinc additions are made with aluminium as they have beneficial effects on mechanical properties. Consequently ternary alloys are often employed containing 6–9·5% aluminium and 0·5–2% zinc.

Aluminium/zinc additions raised proof strengths of castings to around 8 kg/mm^2 with ultimate strengths around 20 kg/mm^2. In wrought products, proof strengths ranged from 11–22 kg/mm^2 with ultimate strengths in the region of 30 kg/mm^2. These alloys remain today an important type of magnesium alloy for use at ambient temperatures.

Magnesium–aluminium–zinc alloys were produced extensively in Germany in 1914–1918, but they suffered from corrosion problems in wet or moist environments. Corrosion, however, is not a serious problem in magnesium alloys today. As in the case of many other metals in industrial atmospheres, corrosion of magnesium alloys is inhibited by a protective surface film formed by the initial attack. When serious corrosion does occur, it is due to the breakdown of the protective film. In the early commercial alloys film breakdown occurred readily, mainly due to the presence of iron, nickel and other cathodic metals present as impurities. These caused galvanic corrosion on a micro scale, disrupting the protective films and inducing rapid attack. Over the years, steady improvements in refining and melting techniques have greatly reduced impurity levels and brought significant improvement in corrosion resistance. Modern alloys retain their lustre indefinitely in clean, dry conditions; moist or wet atmospheres lead to only superficial corrosion.

Serious corrosion problems of the type once suffered by the early commercial alloys now occur only in aggressive media such as acids or salt sprays. However, it must be remembered that contamination of magnesium surfaces with iron, as could occur in shot blasting, will have serious adverse effects on corrosion resistance.

In 1925 the important discovery was made that additions of manganese to molten magnesium greatly increased the corrosion resistance of the resulting alloys. Manganese thus became an important alloying addition. It was thought for many years that manganese acted directly to improve the corrosion resistance of magnesium, but it is now recognised that its action is to precipitate out deleterious impurities such as iron, thus rendering them harmless. Only about two tenths of a percent of manganese are required to perform this function and manganese has remained an important addition at this level to control corrosion in magnesium alloys. At higher levels of around 0·5 to 1·5%, manganese is also beneficial in improving mechanical properties of wrought products, particularly their resistance to creep.

The discovery of the beneficial effects of manganese brought about a marked expansion in the use of magnesium alloys. The proof strength of alloys containing aluminium, zinc and manganese remained, however, in the region of only 8 kg/mm^2 in castings and 11–22 kg/mm^2 in wrought materials.

In the period 1939–1946, the insistent demands for improved performance in aircraft stimulated a demand for improved magnesium alloys. One of the difficulties encountered with magnesium alloys was that castings tended to have a coarse, columnar grain structure which adversely affected both the quality of the castings and their mechanical properties. Wrought alloys were also susceptible to grain growth unless working processes were very carefully controlled.

During the late 1930s, zirconium additions to magnesium were shown to have a remarkable effect in refining the grain structure and produced a very fine, stable array of equiaxial grains. Only around 0·5% zirconium was needed to produce cast alloys with average grain diameters as small as 2×10^{-2} mm. This can be compared with the grain sizes of non-zirconium containing alloys, which were often several mm long in castings and at least 2×10^{-1} mm even in wrought products. Zirconium additions were also shown to double the proof strengths of the early casting alloys and to eliminate most of the problems associated with microporosity. Magnesium–zirconium alloys were thus of outstanding interest in promoting strength and quality in castings; furthermore, their fine grain size greatly facilitated ease of working by forging, rolling and extrusion. Zirconium-bearing alloys were destined to become a major factor in the technological development of magnesium.

Because of the great disparities in the melting points of zirconium and magnesium, the problems of producing the alloys on a commercial scale were considerable; they were apparently considered insuperable in wartime Germany. In the U.K., however, persistent efforts to develop a commercial process eventually succeeded, using reducible zirconium compounds. Casting alloys were developed with proof strengths in the range 13–17 kg/mm^2, ultimate strengths being readily obtainable in the range 24–28 kg/mm^2. Wrought alloys with ultimate strengths up to 36 kg/mm^2 were also produced.

Aluminium and manganese are precipitated from liquid magnesium by the addition of zirconium, and are not used, therefore, in zirconium bearing alloys. Zirconium, like manganese, also precipitates iron and acts like manganese in improving corrosion behaviour. Zinc remains in solution in the presence of zirconium and it is thus possible to combine the strengthening action of zinc with the grain refining capabilities of zirconium to form a very useful ternary alloy.

The action of zirconium in developing a fine grain size has been the subject of much interest. It is clearly associated with the presence of zirconium-rich particles produced in freezing, which anchor grain boundaries and prevent significant grain growth. The anchoring action of these particles is shown in Fig. 2, which is taken from a creep specimen of magnesium–0·5% zirconium alloy.

Zirconium-bearing alloys are usually employed when high-strength, high integrity components are required. Because of the cost of zirconium, however, they are more expensive than the standard aluminium/zinc-containing alloys which are still widely used for many magnesium applications.

For high temperature operation, where creep is an important factor, thorium or rare earth additions are beneficial, grain size control being effected by the use of zirconium. Where low proof strengths can be accepted in order to improve resistance to creep, a coarse grain size is often beneficial. The 1·5% manganese alloy, for instance, is heat-treated to produce grain growth in components requiring good creep resistance for nuclear applications.

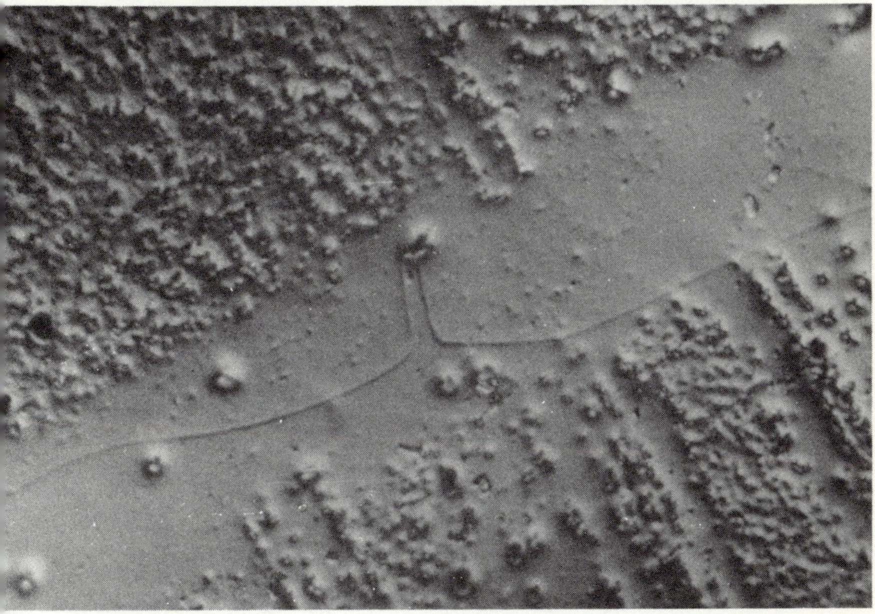

Fig. 2. Anchoring of grain boundary by a zirconium-rich particle. ×500.

For maximum creep resistance, alloys made by powder technology, in which powder is compacted by forging or extrusion, are remarkably effective. Surface oxidation of powder leads to an oxygen content in the material of around 0·1%, and this is more potent in increasing creep resistance than any other alloying element.

Creep resistance is discussed more thoroughly in Section 5.

Beryllium additions of only 0·01% reduce high temperature oxidation rates, raise ignition temperatures and thus ease the problems of handling liquid metal. Beryllium is sometimes added to alloys, therefore, particularly when they are to be used in high-temperature applications.

Table 1

ALLOY TYPE	CHEMICAL COMPOSITION (%) Al	Zn	Mn	Zr	Th	Rare earths	MECHANICAL PROPERTIES 0·1% Proof (kg/mm^2)	UTS Kg/mm^2	Elongation
Casting alloys									
Standard aluminium alloy	6–9·5	0·5–2	0·3				6–11	14–25	4–7
High-temperature alloy		0–2		0·6		1·7–2·5	16–20	25–28	2–8
High-strength zirconium alloy		4·5		0·7			13–17	24–28	5–12
Creep resisting		0–4		0·7	2–3		8–14	20–25	5–8
Wrought alloys									
Standard aluminium alloy	3–8	0·5–1	0·3				11–22	23–31	8–10
Manganese bearing, corrosion resistant			1·5				6–17	20–28	5–10
High-strength zirconium bearing		3		0·6			16–25	25–36	10–15
Creep resisting		0·5		0·6	0·7		9–13	20–24	5–10

The more common casting and wrought alloys used today are listed in Table 1, which shows typical values for 0·1% proof strength, ultimate tensile strength and ductility. The values are intended only as a general guide, as they can be varied by such factors as heat treatment, grain size and structure. There are many more alloys available commercially than those shown in Table 1, which is intended to illustrate only the broad groups of the different alloy types available.

4. Casting, Hot Working and Fabrication

(1) CASTING

The use of magnesium components was pioneered using castings. Today, wrought products comprise an appreciable part of the total magnesium market, but the sale of castings still exceeds that of wrought alloys. The use of castings required the development of special casting techniques, particularly to overcome problems caused by the reaction of molten magnesium with air. The low density and low specific heat of magnesium also had to be taken into account. Castings had to be of fine grain size, have good mechanical properties and be free of flux, porosity or cracks. Complex and often large shapes were required.

Casting alloys were developed to reduce or eliminate the large columnar grain structure of cast pure magnesium, thereby improving strength and ductility. Fluxes were developed which both refined the melt and prevented oxidation of the molten metal. The refining action of these fluxes was brought about by the removal of insoluble non-metallic impurities such as oxides and chlorides from the molten metal. Although the corrosion resistance of magnesium alloys is usually good, it is adversely affected by the presence of chlorides retained in the metal from the electrolytic reduction processes. Since these chlorides are often deliquescent and increase the rate of attack by moisture, their elimination is of great importance. It is equally important that none of the refining flux, itself containing chlorides, is carried over with the molten metal into the casting during pouring.

By varying the flux composition, its degree of fluidity can be varied. During the early stages of melting, very fluid fluxes are used. Very stiff fluxes are particularly suitable for the final pouring operations, as they are easier to separate from the molten metal, thus preventing flux from entering the mould. Modern fluxes consist of mixed chlorides of potassium, barium, magnesium, calcium and sodium. Magnesium and calcium fluorides may be included in the stiffer fluxes. Manganese chloride is also used, not only to protect the melt but also to control the amount of manganese in the alloy.

Magnesium can be cast into silica sand moulds or into permanent moulds using gravity die or pressure die casting methods. Steel is used for the permanent moulds. With any metal of reasonably low melting point, pressure die casting is an extremely economical method for making a large number of identical parts. In this process a fixed quantity of liquid metal is poured into a chamber and is then compressed by an hydraulically operated pressure plunger. This forces the metal into the die, where it solidifies rapidly by cooling through the die walls. The die is opened, the casting removed and the operation repeated. The process is readily automated and is particularly suitable for magnesium in producing low-cost components, largely because of the low specific heat and low density of the metal. These properties give magnesium a particularly low volume heat capacity, thus leading to easy cooling of the mould and promoting very rapid production cycles. In addition, magnesium castings do not stick to the dies, which therefore have a very long life.

Until the discovery of the grain refining properties of zirconium, magnesium alloys with aluminium, zinc and manganese formed the mainstay of the magnesium foundries as they could be produced with reasonable grain size and mechanical properties. The development of the zirconium-bearing alloys enabled fine grained castings to be readily obtained. Casting alloys containing zinc and zirconium have excellent properties and are generally free of micro-porosity problems. The alloys came into widespread use, and they quickly became as important as the long established magnesium–aluminium based alloys. In general, zirconium alloys are now used in both the cast and wrought forms in the aircraft and missile fields or where optimization of properties is more important than cost. For more conventional applications, the cheaper magnesium-aluminium alloys are still widely used.

Typical cast alloy compositions are shown in Table 1 with their mechanical properties.

(2) FORMING

Since World War II, there has been a marked growth in the use of wrought magnesium alloys at the expense of castings. This has arisen from the demand for both improved mechanical properties and for large quantities of sheet and strip. The availability of improved cast ingot is

Fig. 3. Fuel element assembly for carbon-dioxide-cooled, uranium-fuelled reactor, showing impact extruded magnesium alloy can, extruded splitters and forged braces. End cap weld is by automatic tungsten inert gas. Splitter and brace assembly is spot welded.

another factor. As ingot has become more homogeneous and finer in grain size, many of the difficulties associated with mechanical working have been eliminated.

The hexagonal nature of the magnesium crystals limits the amount of deformation which can be tolerated at room temperature, but magnesium alloys can be hot worked readily by all the usual forming processes. During the 1950s and 1960s, working speeds increased markedly and a wide variety of complex shapes are now made by extrusion and forging.

Extrusion is usually conducted at billet temperatures of 300–400°C at speeds between 5 and 200 feet per minute. Magnesium alloys extrude readily, and less control on homogeneity and grain size is required than is demanded for the more exacting processes of rolling and forging. Impact extrusion can also be employed for symmetrically shaped tubular products requiring thin walls of irregular profile. Such a component is used in the fuel element assembly shown in Fig. 3. This shows an impact extruded finned hollow can which contains a uranium fuel rod for use in nuclear reactors. Also shown in this illustration are extruded magnesium alloy T-shaped splitters and forged braces supporting them.

Rolling is conducted at temperatures up to 500°C using the biggest reductions possible at each pass in order to reduce the number of reheats. Thus, billets weighing half a ton can be reduced to $\frac{3}{4}$ inch sheet without reheating, provided the cast ingots are of fine grain size as is the case with zirconium bearing alloys.

Magnesium is easy to forge provided the grain structure is fine, when drop forging can be employed. With coarser grained starting stock, slow acting hydraulic presses are used. Figure 4 shows a forged fuel element spider, illustrating the complexity of shape that can be realised. Magnesium can also be readily deep drawn, spun and stretch formed.

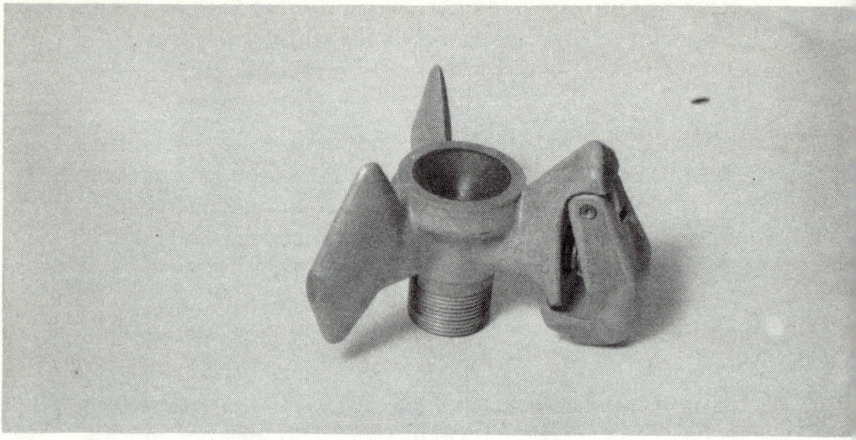

Fig. 4. Forged fuel element spring arm spider.

Magnesium alloys are readily weldable by inert gas shielded tungsten arc processes, a method which has been fully automated in the welding of end caps into fuel element cans. Such a weld can be seen in Fig. 3. Consumable electrode inert gas processes and resistance welding are also widely used effectively. Several resistance welds can be seen in the fuel element shown in Fig. 3; resistance welding is highly useful for economic fabrication of magnesium parts. Gas welding of magnesium is now largely obsolescent, mainly because the corrosive nature of the welding fluxes required can lead to heavy corrosion of the welds.

Magnesium alloys are noted for their ease of machining, and metal cutting is conducted at very high speeds with deep cuts and high feed rates. Surface finish is excellent and machining is a highly economic method of component production.

5. Creep Resistance

The long-established magnesium alloys containing aluminium and zinc are not very suitable for use at temperatures above 100°C as their mechanical properties deteriorate rapidly as the temperature is raised. However, additions of the rare earth metals improve the high temperature properties and such additions are usually used in conjunction with zirconium to control grain size. A typical casting alloy developed for high temperature use contains 2·5% rare earths, 2% zinc and 0·7% zirconium. Creep data on this alloy at 250°C are shown in Fig. 5. The castability of the alloy is outstanding and for this reason it is sometimes used even for room temperature application.

RE additions

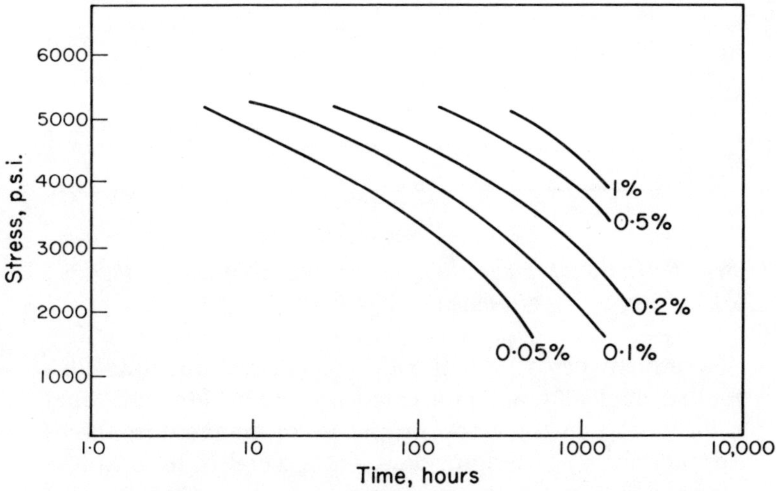

Fig. 5. Creep of magnesium alloy containing 2·5% rare earths, 2% zinc, 0·7% zirconium at 250°C.[4]

For higher temperature service, thorium additions are beneficial and casting alloys typically containing 2·7% thorium, 2·2% zinc and 0·7% zirconium are employed. These have better properties than the rare earth containing alloys at 250°C (see Fig. 6) and are used up to 350°C (see Fig. 7). The ratios of the thorium to zinc contents are critical in determining the long-term creep properties of these alloys. This effect can be seen in Fig. 8, which also illustrates the dangers inherent in drawing conclusions from one-day creep tests. The strengthening effects of the rare earths and thorium are exploited in producing wrought alloys for use at elevated temperatures. Manganese has also been used successfully as an alloying addition to promote high temperature strength.

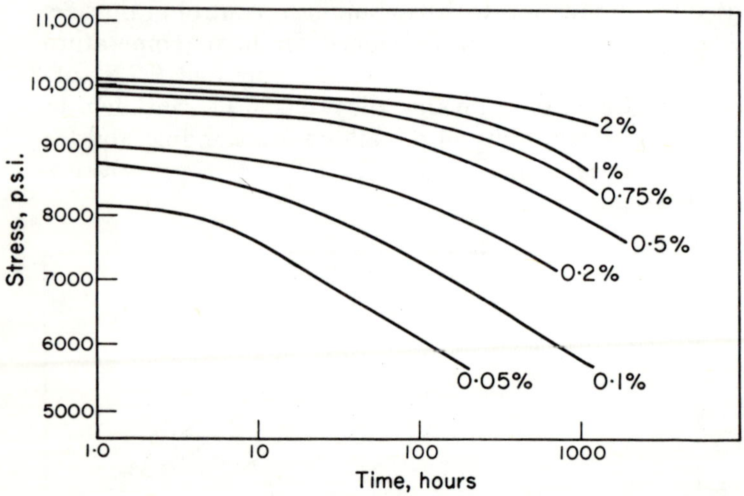

Fig. 6. Creep of magnesium alloy containing 2·7% thorium, 2·2% zinc, 0·7% zirconium at 250°C.(4)

The advent of carbon dioxide cooled, natural uranium-fuelled nuclear reactors created a demand for additional information on the creep properties of magnesium alloys. In fuel-element design, data were needed to optimise components such as end caps and secondary support components such as splitters, braces and spring-arm spiders which were required to operate at temperatures up to 450°C.

A number of alloys have been examined extensively for such applications in the U.K. nuclear programme. Rare earth additions proved unsuitable, because the alloys had

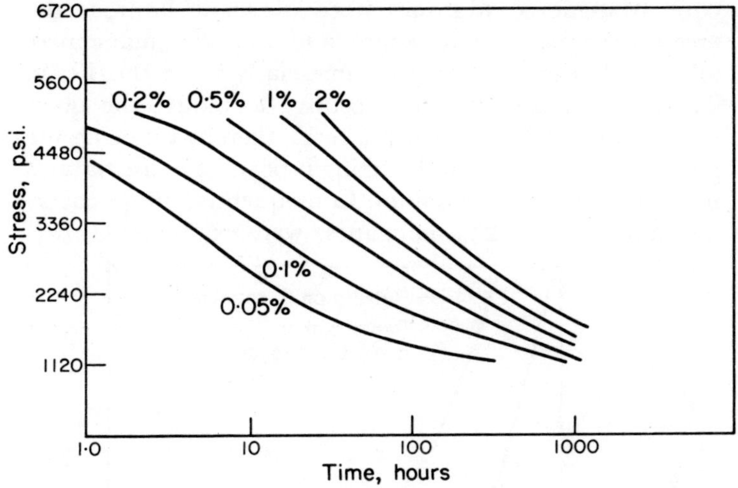

Fig. 7. Creep of magnesium alloy containing 2·7% thorium, 2·2% zinc, 0·7% zirconium at 350°C.[4]

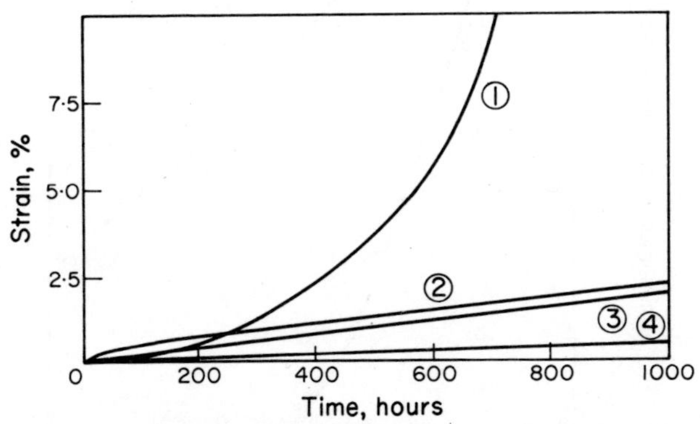

1 Mg — 0·7% Zr — 3% Th
2 Mg — 0·7% Zr — 3% Th — 3·4% Zn
3 Mg — 0·7% Zr — 3% Th — 1·2% Zn
4 Mg — 0·7% Zr — 3% Th — 2·5% Zn

Fig. 8. Effect of zinc additions on the creep resistance of cast thorium-bearing alloys, 320°C, 3360 p.s.i.

inadequate resistance to attack by the carbon dioxide coolant. Thorium was also an unacceptable alloying element as it could undergo nuclear fission in the reactor core. Manganese additions were known to improve the creep strength of magnesium and a 1–2% manganese alloy had been available commercially since the 1930s. There was a considerable incentive to reduce the manganese content for nuclear purposes, thereby reducing the neutron cross-section and making it possible to use natural instead of enriched uranium. Consequently, alloys covering the range 0·5–2% manganese were examined.

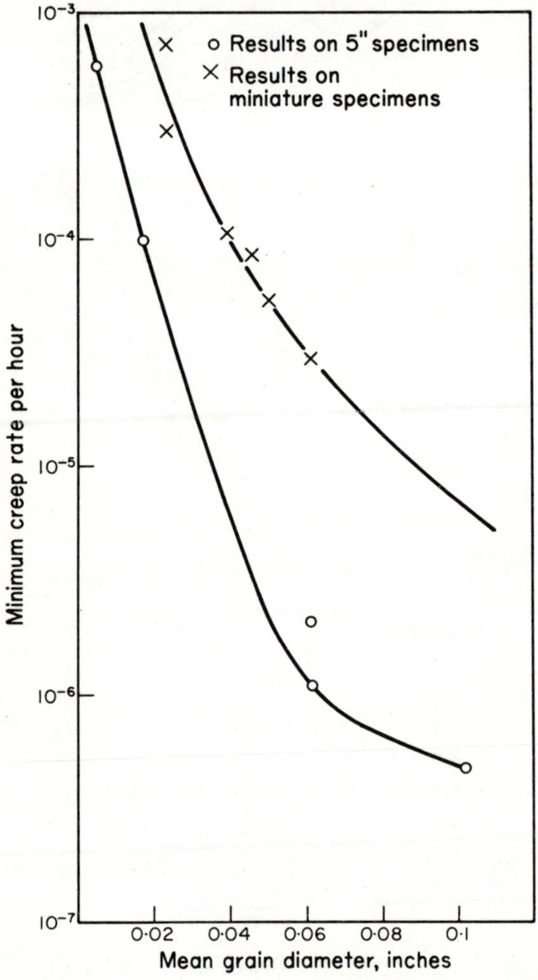

Fig. 9. The effect of grain size on the creep resistance of a magnesium–0·5/1·0% manganese alloy at 450°C and 224 p.s.i.

Grain size was shown to play a very large part in determining creep properties, a grain-coarsening treatment improving the creep strength significantly but lowering ductilities. Grain size effects were investigated principally on the 0·5–1% manganese alloys; the relationship between grain size and minimum creep rate in tests at 450°C and 224 p.s.i. is shown in Fig. 9. The creep properties of this alloy in the coarse grained condition (2·5 mm average grain diameter) are shown in Fig. 10. Examination of the effect of manganese content on the creep properties of this range of alloys has shown that at constant grain size, the strengths at 400 and 450°C are raised as the manganese content is increased, probably as a result of increased precipitation hardening. In practice, both the 0·5–1% manganese alloys and the 1·5–2% manganese alloys are used for various components, depending on the stress, temperature and creep strains allowable.

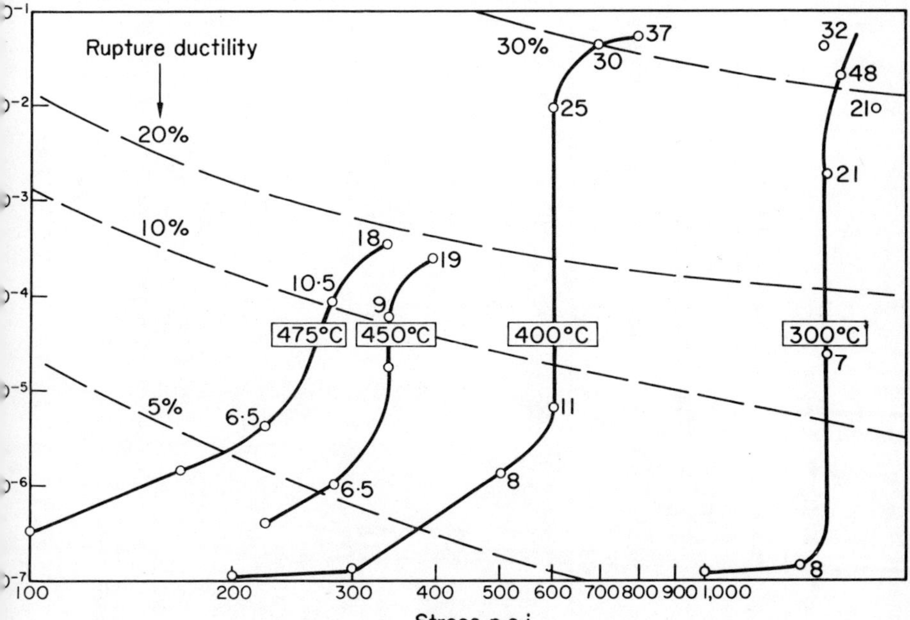

Fig. 10. Creep of a 0·5/1·0% manganese alloy heat-treated to a grain size of 2·5 mm average grain diameter. The figures by the data points are values of rupture ductility.

Table 2 Creep Tests at 400°C and 200 psi.

NO.	SPECIMEN TREATMENT	GRAIN SIZE (in. $\times 10^3$) after treatment	ZIRCONIUM CONTENT After treatment		ZIRCONIUM CONTENT After creep test		HYDROGEN CONTENT (ppm)		CREEP TEST (hrs.)	CREEP STRAIN (%)	CREEP STRAIN AFTER 2000 h (%)	FINAL STATE
			(sol. %)	(insol. %)	(sol. %)	(insol. %)	After treatment	After creep test				
1	As extruded	1–2	0·47	0·04	0·47	0·04	5–14	17	1 443	125	—	Ruptured
2	96 h in CO_2 with 50–300 ppm water at 600°C	22	0·49	0·02	0·02	0·41	5–7	160	16 656	10	1·88	Ruptured
3	96 h in CO_2 with 1% water at 600°C	11	0·07	0·3	0·01	0·42	—	107	3 338	0·1	0·085	Discontinued
4	96 h in CO_2 followed by 16 h in hydrogen at 600°C	14	0·02	0·53	0·01	0·55	150	153	4 190	0·33	0·236	Discontinued
5	16 h in hydrogen followed by 96 h in CO_2 at 600°C	13	0·04	0·45	0·02	0·50	—	146	1 200	0·47	0·6*	Discontinued
6	96 h in hydrogen at 600°C	5	0·01	0·48	0·01	0·45	—	135	2 269	14	9·3	Ruptured

* Extrapolated value

The magnesium–0·5% zirconium alloy is particularly interesting for nuclear applications because of its low cross-section to thermal neutrons. This alloy was first examined for possible use as a fuel can, for which outstanding creep ductility was required rather than creep strength. It was soon discovered, however, that its resistance to creep could be altered by several orders of magnitude by heat treatment. This is illustrated in Table 2 and Fig. 11. Factors effecting creep strength are grain size and the distribution of zirconium phases, particularly zirconium hydride. In the as-extruded condition without heat treatment the alloy is fine grained, very weak but highly ductile in creep, with the zirconium largely in solution. By homogenizing the alloy, promoting grain growth and then precipitating the zirconium as hydride, a structure very resistant to creep deformation can be produced. The extent of this improvement can be seen in Table 2, which shows that at 400°C and 200 p.s.i., the as-extruded alloy ruptured in 1443 hours with 125% strain.

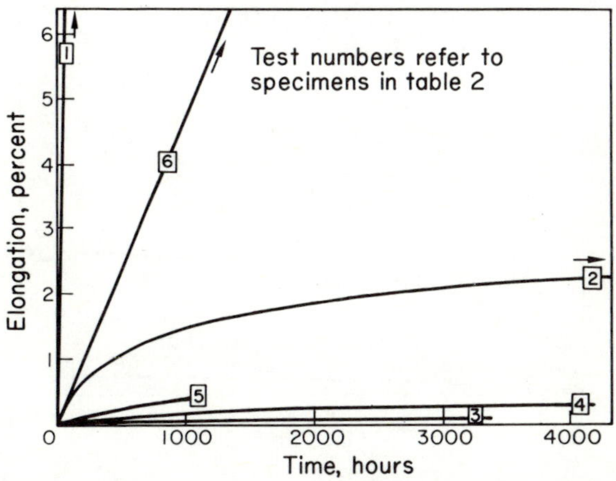

Fig. 11. Magnesium–0·5% zirconium alloy in various heat-treated states, creep tested in air at 400°C with a stress of 200 p.s.i. The test numbers refer to specimens in Table 2.

After homogenisation and hydriding treatments, under the same conditions of test the alloy deformed only 0·33% in 4190 hours when the test was discontinued. This heat treatment inevitably entails some loss in ductility, but structures of intermediate strength can be readily obtained by variation in heat treatment procedures. This is a rare example of the effective use of hydrogen as an alloying element in metals.

If the magnesium–zirconium alloy is not hydrided before use, it can react slowly with moisture in the environment at temperatures above 400°C to form zirconium hydride, the reaction extending throughout the mass of the alloy. The reaction modifies the properties of the alloy during service, slowly strengthening it at the expense of ductility. The pre-hydrided alloy, on the other hand, is stable; the hydriding process has been taken to completion before use. These effects are shown in Fig. 12.

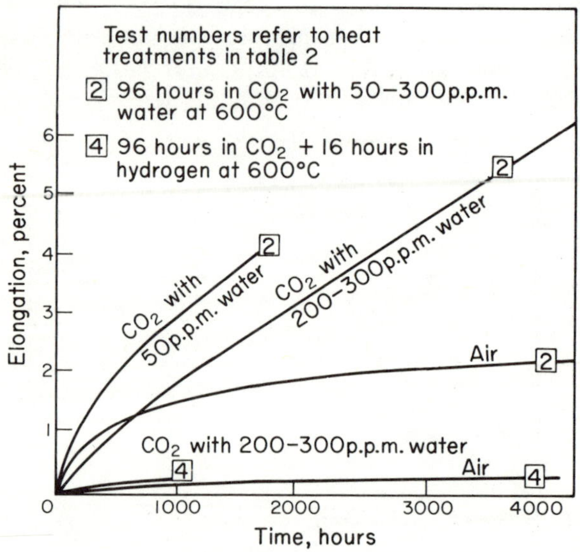

Fig. 12. The effect of creep-test atmosphere on the strength of a magnesium–0·5% zirconium alloy at 400°C and 200 p.s.i.

In France, this alloy has found use in the as-extruded condition as a fuel element can. In the U.K., it has been used in the strengthened, heat-treated condition as fuel element support components such as splitters, which also improve heat transfer characteristics. A heat treatment of

96 hours at 600°C is employed for such components prior to service.

Since optimisation of fuel element design required magnesium alloys to operate at their limits of creep strength, investigations of conventional wrought alloys were supplemented by an appraisal of dispersion hardened alloys made by powder metallurgical techniques. Initially, the alloys were produced by compaction and extrusion of raspings, a term given to particles resembling lathe turnings made by scratching the surface of a solid block of material with a suitable cutting tool. Raspings were chosen instead of the more conventional powder form because they were readily available commercially for use in uranium reduction, and because their size enables them to be handled without the need for elaborate fire precautions. Investigations were also made into the properties of alloys produced from powder obtained by filing the

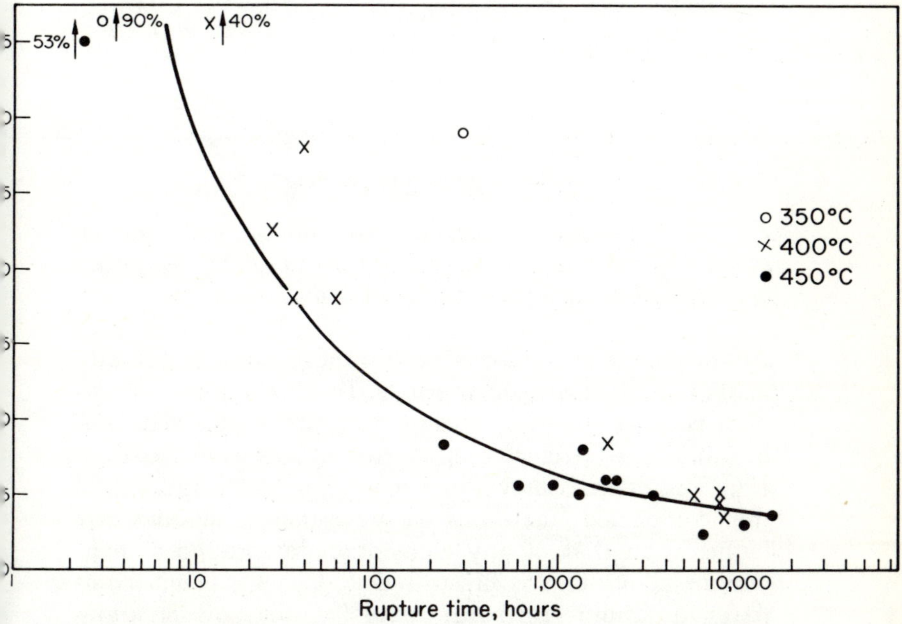

Fig. 13. Rupture strain against time to rupture at 350, 400 and 450°C for a powder compacted and extruded magnesium–0·5% zirconium alloy.

surface of cast material with an abrasive endless belt. This powder was considerably finer and more uniform in shape than the raspings.

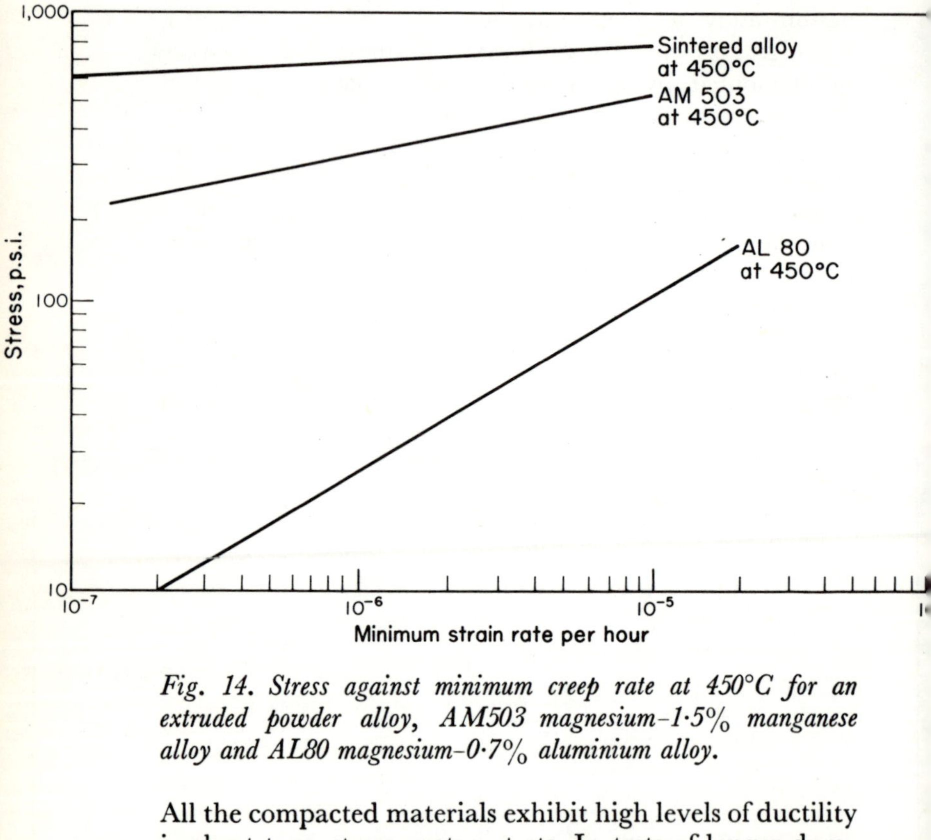

Fig. 14. Stress against minimum creep rate at 450°C for an extruded powder alloy, AM503 magnesium–1·5% manganese alloy and AL80 magnesium–0·7% aluminium alloy.

All the compacted materials exhibit high levels of ductility in short term stress rupture tests. In tests of longer duration, rupture ductilities below 10% were observed, and in some tests tertiary creep started at strains of less than 1%. This effect is shown in Fig. 13. The creep strength of the compacted materials is nevertheless significantly higher than that of any conventionally produced magnesium alloy. This is illustrated in Fig. 14, which compares minimum creep rates at 450°C for powder alloys with cast and extruded magnesium–manganese and magnesium–aluminium alloys. It can be seen that the compacted magnesium alloy has 3–4 times the stress-bearing capacity of the best cast and worked magnesium alloys. Despite these improvements in properties, it has not proved necessary to date to use these alloys, the cheaper, conventionally produced components proving adequate for all requirements.

6. Fatigue

Rotating bend fatigue tests conducted in air on small samples of magnesium alloys show the usual form of relationship between fatigue stress and life shown in Fig. 15.

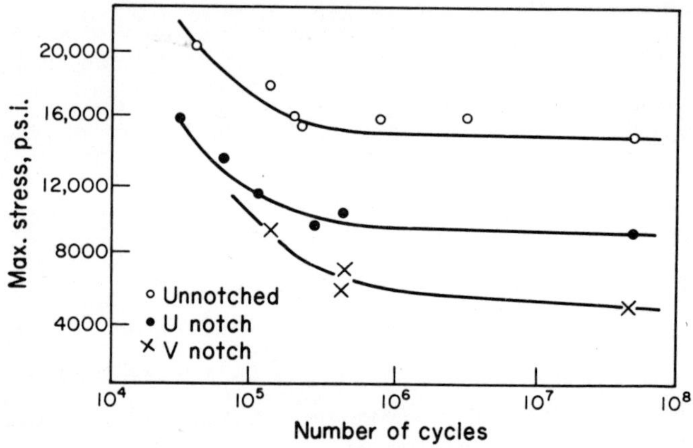

Fig. 15. Fatigue resistance of a magnesium alloy containing 1·7% rare earths, 0·6% zirconium and 2·5% silver.(4)

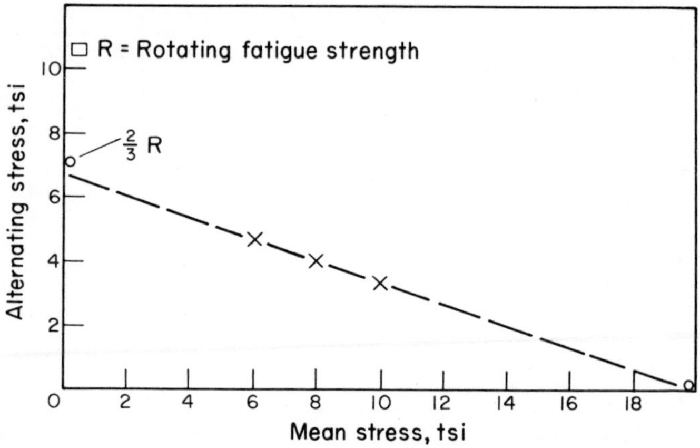

Fig. 16. Effect of mean level of stress on push–pull fatigue strength for a magnesium alloy containing 2·7% rare earths, 2·2% zinc and 0·7% zirconium.(4)

It would appear, however, that a non-oxidising environment is required to produce a true fatigue limit. Nevertheless, the rate of fall in the fatigue strength in air is extremely slight after 10^6 cycles have been exceeded. The fatigue stress for failure in bending in 5×10^7 cycles ranges between 0·3 and 0·5 of the ultimate tensile strength; this is not a dissimilar figure to that found in other materials. As with other alloys, the fatigue strength goes down as the specimen diameter is increased. Tests in push–pull instead of bending also lower the fatigue limit, generally to about two-thirds of the bending fatigue values. As is generally the case, fatigue strengths are also lowered by an increase in the level of the steady stress. These effects are shown in Fig. 16.

Water and salt solutions seriously impair fatigue strength. The presence of water alone reduces fatigue strength by a factor of approximately two, whereas a 3% salt solution can reduce strengths by between three and ten times in terms of stress. This reduction in strength is clearly brought about by a corrosion fatigue process, which possibly also accounts for the lack of a true fatigue limit even in air. A good fatigue strength in the presence of chlorides therefore depends on the provision of surface treatments to eliminate corrosive attack.

7. Flammability

Magnesium is often thought of as a highly flammable material which will burn readily on heating. This is not the case. Magnesium components will burn in air only if they are heated to temperatures approaching their melting point, and some alloys must be melted before they will burn. Magnesium–zirconium alloys, for instance, will ignite at 625°C in moist air. Small additions of beryllium (in the region of 0·01%), however, raise the ignition temperature to above the melting point and even prevent oxidation of the liquid metal and reduce the tendency to burn.

As with many other materials, the risks of ignition increase as the metal is reduced to a finely divided form. Even so, coarse chips produced during machining cannot usually be ignited. Fine swarf may burn under certain conditions, for example when intense frictional heat is developed by the usc of blunt cutting tools. Swarf fires, whilst uncommon, do occur; they are not unduly dangerous, however, because although they emit a very intense light, they burn slowly and evenly. If water is used in an attempt to extinguish the fire, hydrogen generation stimulates the blaze. It is usual to have buckets of cast iron chippings or dry sand available to extinguish any fires which occur during machining.

Magnesium dust is explosive if heated, and a spark produced during grinding may be sufficient to cause rapid ignition. Dust must be handled with extreme care; it must be removed continually from the workshop and drenched under water. When it is necessary to dry grind, powder must be extracted immediately from the grinder by forced air and then immersed in water. Such precautions are not peculiar to magnesium; aluminium, zirconium and other powders are also explosive and must be treated with care.

Flammability of magnesium is not an important factor in

the engineering utilisation of the metal. Alloys of magnesium are used at temperatures up to 500°C in the cores of nuclear reactors with perfect safety, albeit in an environment of carbon dioxide.

8. Corrosion and Corrosion Protection

The early magnesium alloys were susceptible to heavy corrosion in moist conditions. This was attributable to impurities, particularly iron and nickel, which were present in the magnesium. These acted as microscopic cathodes which destroyed the protective oxide film on the metal surface. Both manganese and zirconium additions to molten magnesium precipitate iron, nickel and other harmful impurities from solution, thus greatly improving the corrosion resistance of modern alloys.

Table 3. Results of 2½-year exposure tests

MATERIAL	CORROSION RATE (thou/annum)	% LOSS TENSILE STRENGTH (after 2·5 years)
Marine atmosphere, Kure Beach, North Carolina, U.S.A.		
Aluminium	0·06	2·5
Magnesium	0·70	7·4
Mild steel	5·91	75·4
Industrial atmosphere, Madison, Illinois, U.S.A.		
Aluminium	0·08	1·5
Magnesium	1·09	11·2
Mild steel	1·00	11·9
Rural atmosphere, Near Midland, Michigan, U.S.A.		
Aluminium	0·005	0·4
Magnesium	0·53	5·9
Mild steel	0·59	7·5

Unprotected metal, as made today, is very resistant to atmospheric attack; it is not as resistant to corrosion as aluminium, but it is better than mild steel. The behaviour of these materials in various atmospheres is compared in Table 3.[3]

Newly machined magnesium will retain its metallic lustre for a number of days, but in a damp environment it will gradually become dull and light grey in appearance. The main corrosion product is the hydroxide $Mg(OH)_2$. The overall rate of attack in normal industrial atmospheres is negligible. In more severe environments, such as heavy salt sprays, pitting and general corrosion may occur; magnesium requires protection from chloride ions for prolonged survival. Magnesium is also attacked extensively by many acids, but is resistant to corrosion in alkalis.

Various forms of protective treatment have been devised for magnesium, for use particularly in marine environments. Coatings include chromating, anodising, surface sealing with spray resins, electroplating and painting. Chrome acid pickle is used to produce a very thin protective film which provides useful protection during storage, and serves also as a paint base. The anodised layers are thicker and harder, but a paint finish is desirable in the worst environments. Anodised layers are, however, highly resistant to wear.

In assembly, care must be taken to avoid galvanic corrosion when magnesium is used with other metals. This is best achieved by using non-absorbent tapes or sealing compounds in joints to eliminate contact between the metals. Contact with wood must also be avoided in wet atmospheres, as water tends to leach out from the timber natural acids, which then attack magnesium.

9. Costs

Aluminium is commonly the principal competitor with magnesium in many applications, and a comparative cost analysis is shown in Fig. 17. This includes prices for both primary and recycled aluminium compared with magnesium. There is an appreciable market in secondary aluminium, but there is at present only a negligible market in secondary magnesium. According to the U.S. Bureau of Mines, for instance, consumption of magnesium-base scrap metal is only 5% of the total used. This is presumably accounted for in part by the high proportion of magnesium used destructively for chemical applications or for alloying with aluminium.

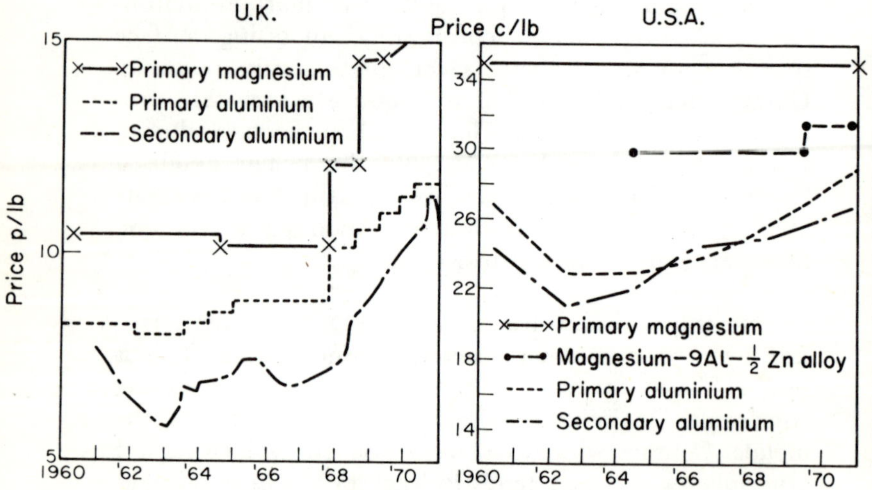

Fig. 17. Price of magnesium and aluminium ingot, U.K. and U.S.A., 1966–1970.(2)

The ratio of selling prices of magnesium to secondary aluminium for U.K., U.S.A. and Germany is shown in Fig. 18. The price ratio giving equal cost per volume is 1·6 to 1. In the future, the ready availability of magnesium compounds and improvements in extraction processes (e.g. chlorine recovery) will probably reduce the cost differential between magnesium and aluminium. It is predicted[2] that even on a weight basis the ratio of production of magnesium to aluminium will fall to 0·9/1·0 during the 1970s.

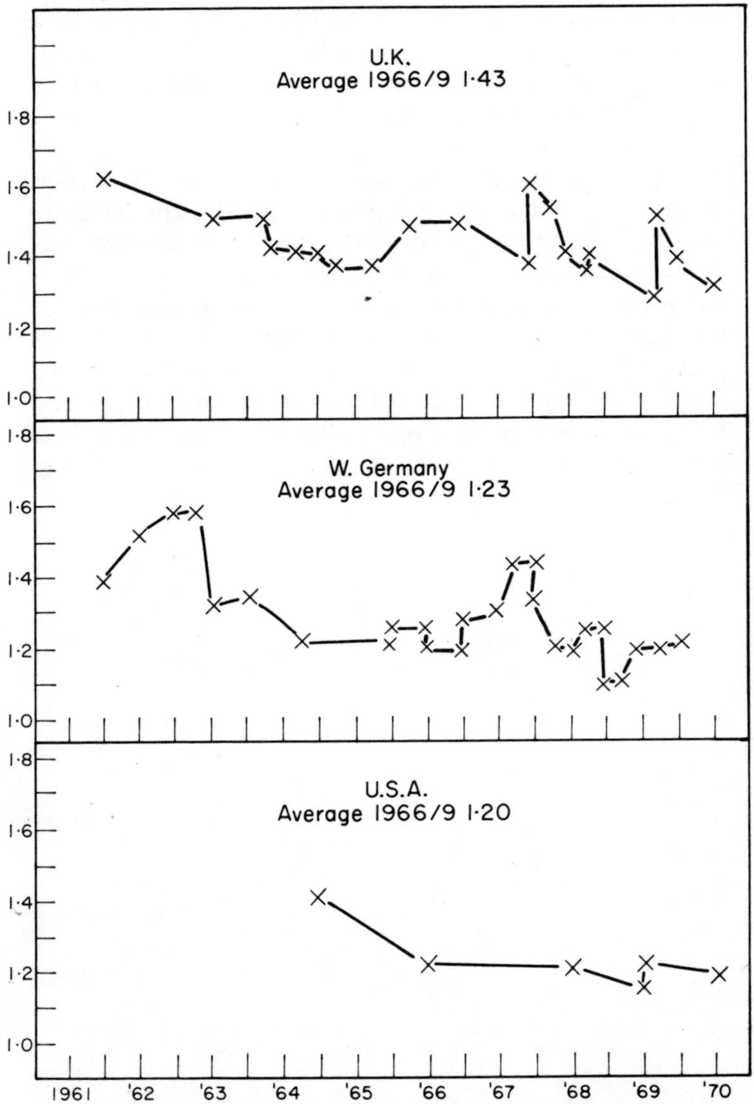

Fig. 18. The ratio of selling prices by weight of magnesium to secondary aluminium for U.K., U.S.A. and West Germany, 1961–1970.[2]

To fully utilise this reduction in costs, it will be necessary to improve conversion efficiencies. These are at present poor compared with aluminium, because of the need to use fluxes in melting and the greater loss of metal. Efficient conversion can often be achieved in processes such as pressure die casting used on a large quantity basis, and it is here that the magnesium market is most likely to expand in engineering components.

The price of magnesium ingot is shown in Fig. 17 as around 15p (36 cents) per pound, but the cost of magnesium components is of course higher than this. The actual cost is affected by many factors, such as the complexity of the design, size, wall thickness, quality and the number of pieces required. As a general indication, however, simple die castings heavier than one-third of a pound cost around 50p (120 cents) per pound provided more than 1000 pieces are made. For smaller numbers and lighter weights, the cost may be five or six times greater than this.

10. Applications

(1) CHEMICAL USES

A review of the applications of magnesium is to be found in *Principles of Magnesium Technology*,[4] where it is estimated that about half the magnesium produced is used for chemical rather than for constructional purposes. A breakdown of the utilisation of magnesium into chemical and structural uses in the U.K., U.S. and Germany is shown in Fig. 19.[2]

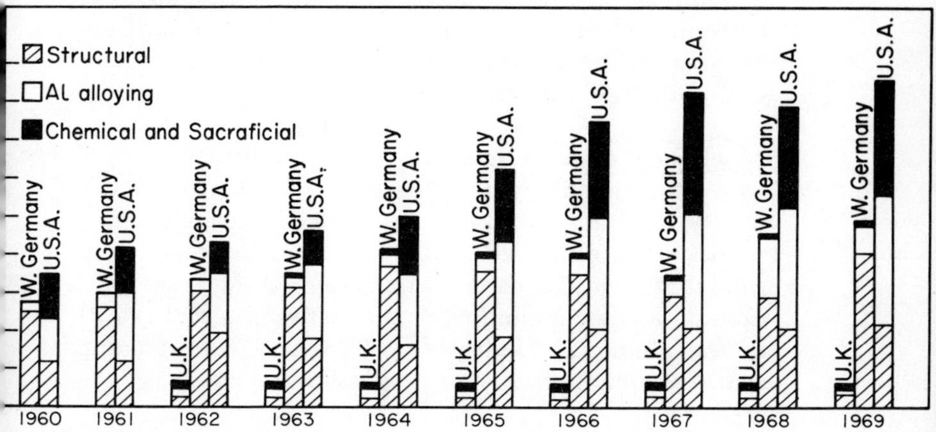

Fig. 19. Consumption of magnesium in U.K., U.S.A. and West Germany 1960–1969.[2]

Many of the chemical applications arise from the high affinity of magnesium for oxygen. It is widely used for the reduction of the oxides of other elements in many metal extraction processes, particularly in the production of uranium, titanium and zirconium. It is also used in pyrotechnics, where it is burned to produce attractive fireworks with strong, slow-burning white or coloured light. Magnesium flares are used for military purposes in Very lights and magnesium powder can be used for showing the path of projectiles as in tracer bullets.

Magnesium is used for reserve cell battery plates. The reserve cell is assembled dry, in which condition it can be kept indefinitely. To activate the cell, it is only necessary to immerse it in water after which large voltages can be developed for a short time. The low density of magnesium makes it very satisfactory for use as anodes in these cells, which are used often in situations where the saving of weight is an important factor, for example in the powering of sea rescue devices. Magnesium anodes can also be used in dry cells.

The very active electrode potential of magnesium makes it highly suitable as a sacrificial anode and it will protect most of the commonly used engineering materials from corrosion when it is connected with them electrically. Magnesium anodes are utilised in the protection of ships' hulls, underground steel pipe-lines and storage tanks.

Much magnesium is used as a minor constituent in alloys with other metals; with aluminium, in particular, it is widely used for promoting improved mechanical properties. It is estimated that 30–35% of the magnesium produced today is used in this way. On average, aluminium alloys contain 0·85% magnesium, so that growth in the aluminium industry will lead to corresponding growth in requirements for magnesium.

Magnesium is used in many organic chemical processes as a Grignard reagent, in which magnesium is bound into compounds of the RMgX type, where R is the organic radical and X is a halogen. These organomagnesium compounds are used, for instance, as intermediates in the conversion of aldehydes and ketones to alcohols.

(2) AIRCRAFT

The engineering applications of magnesium arise largely from its light weight, a feature which makes it attractive for many components used in transport. A detailed analysis of consumption in the U.S. in the period 1955–1958 is given in Table 4.[3]

Table 4. Utilisation of magnesium in the U.S., 1955–1958

	TONS				PERCENTAGE OF TOTAL			
	1955	1956	1957	1958	1955	1956	1957	1958
Structural Uses								
Aircraft and missiles	11,500	11,300	10,660	9,774	17·7	16·3	19·5	22·0
Consumer products	2,050	2,250	1,650	2,300	3·1	3·2	3·0	5·2
Electrical & electronic equipment	700	950	1,100	1,460	1·1	1·4	2·0	3·3
Machinery & tools	3,150	3,900	2,830	2,290	4·8	5·6	5·2	5·2
Material-handling equipment	1,100	1,150	1,150	890	1·7	1·7	2·1	2·0
Surface vehicles	1,250	1,050	1,580	791	2·0	1·5	2·9	1·8
Graphic arts	450	500	430	250	0·7	0·7	0·8	0·6
Miscellaneous	250	400	970	670	0·4	0·6	1·8	1·5
Total structural uses	20,450	21,500	20,370	18,425	31·5	31·0	37·3	41·6
Non-structural Uses								
Electrochemical & chemical	6,800	7,300	5,590	5,036	10·5	10·5	10·2	11·5
Alloying with aluminium	14,000	15,550	12,750	11,415	21·6	22·5	23·4	25·7
Reduction of other elements	14,750	20,100	13,750	7,940	22·8	29·0	24·7	17·9
Powder	900	1,250	850	420	1·4	1·8	1·6	0·9
Miscellaneous	50	50	50	40	0·1	0·1	0·1	0·1
Total non-structural uses	36,500	44,250	32,990	24,851	56·4	63·9	60·0	56·1
Export (all forms)	7,850	3,500	1,500	1,052	12·1	5·1	2·7	2·3
Total	64,800	69,250	54,860	44,328				

Extensive use of magnesium components is made in aircraft components, including airframes, engines and landing gear, and this market has probably been the biggest single factor in promoting magnesium technology. Both castings and wrought products are used, for the most part using zirconium bearing alloys. Reference 4 quotes the "Comet", "Britannia" and "Trident" as examples of aircraft which contain several hundred magnesium castings. Major structural pieces include cockpit canopies and windscreen supports, gear cases and undercarriage legs. Wrought components include fuel tanks, air ducts, tubular frames for seats, and flooring. Small magnesium pieces are found in many accessories for ventilation, interior partitioning and in navigational aids. Rolled sheet is used extensively for aircraft skins. Millions of cast landing wheels have been used, in which light weight is allied to good fatigue and impact properties.

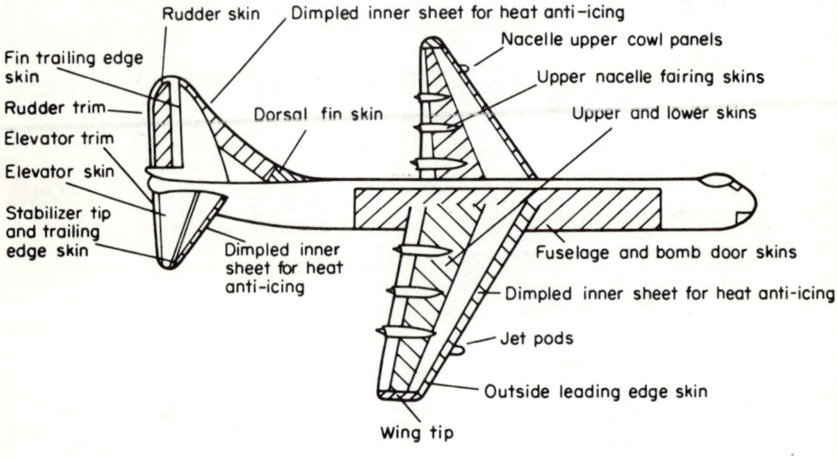

Fig. 20. Magnesium in the B36 bomber.[4]

In the United States, even wider use is made of magnesium in aircraft; the B36 bomber, illustrated in Fig. 20, may be taken as an example.[4] The B36 uses 7500 lb of magnesium sheet and a further 2500 lb of castings, extrusions and forgings as part of the structure of the plane. A further 9000 lb of magnesium is used in auxiliary items, making 19,000 lb in all. Replacement of the magnesium by aluminium would have increased the overall weight of the bomber by as much as $4\frac{1}{2}$ tons.

Developments in cast components in the U.S. include a complete experimental cast wing-section, 16 feet long, weighing 200 lb.

In engines, magnesium is used as air intakes, diffuser and compressor casings and support plates. In the "Dart" turbo-prop engine used on the "Viscount", magnesium castings account for one-fifth of the overall engine weight.

(3) MISSILES AND ROCKETS

Magnesium is used extensively in rockets and missiles for body skins, stiffeners, pump housings, bulkheads, fins and many auxiliary items. These applications are illustrated in Fig. 21.[4] About 2000 lb of magnesium are used in the

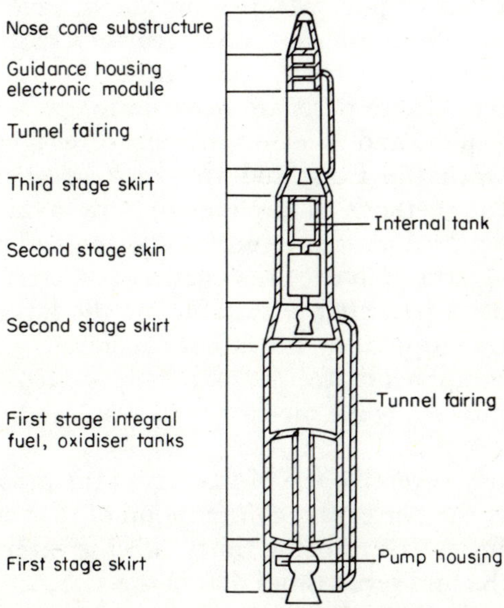

Fig. 21. Hypothetical I.C.B.M. showing area where magnesium can be used to save weight. Actual use in such areas occurs in Titan, Jupiter, Thor or Polaris missiles.[4]

"Titan" Inter-Continental Ballistic Missile. This includes about 40% of the skin. Magnesium is used also in auxiliary equipment which must be easily assembled and where light weight is therefore essential.

"Discoverer" satellites, 19 feet long and 5 feet in diameter, contain over 600 lb of magnesium; this is more than a third of their total weight. The magnesium is used largely in skin and fairings, and in auxiliary equipment used in cameras, tape recorders, spools and other components. The use of magnesium results in a saving in weight of 25% compared with titanium and even more compared with steel.

(4) MOTOR VEHICLES

Magnesium alloys have been used for many years in motor vehicles. Typically, a 55-lb casting forms the crankcase of a bus engine. Magnesium alloys have also been used for clutch housings, sumps, gear-box casings, transmission housings, wheels, instrument panels and body components. These parts are made in magnesium largely to save weight, and to ease problems of weight distribution between the front and the back of vehicles. The economical methods of production now available with magnesium, including pressure die-casting and high-speed machining, also influence the choice of material. Pressure die-casting is particularly suitable for the large quantity production required by the motor car industry, and magnesium components are competitively priced compared with aluminium when die-casting is employed.

Particularly extensive use of magnesium is made in West Germany, the per capita consumption of that country for engineering purposes being thirty times greater than that of the U.K. and seven times that of the U.S.A.[2] Much of this (some 38,000 tons per annum) is used in the manufacture of Volkswagen cars. The specific requirement of VW for a light engine necessitates the use of 44 lb of magnesium in the 1200 cc engine and 50 lb in the 1600 cc engine, largely for the crankcase and transmission housing. The Porsche 917 sports car contains 300 lb of magnesium alloy castings of which the engine contains 155 lb. Pressure die-cast autowheels also account for over 1300 tonnes per year.

Even without the massive contribution of the German motor car industry, the per capita consumption of magnesium in that country would still be higher than that of any other country. This possibly reflects the development of the metal as a strategic commodity in Germany during the 1930s, which has led to its wide acceptance as an engineering material.

(5) NUCLEAR POWER

Magnesium was inevitably chosen as the material for use as fuel element cans in the Calder Hall type of commercial nuclear reactors, because every other metal was unsatisfactory. Fuel-element cans are used to prevent oxidation of the fuel by the coolant gas, to provide efficient heat transfer surfaces and to retain fission products within the can. As natural, un-enriched uranium was chosen for the fuel, however, it was essential to use a can with a very low cross-section to thermal neutrons. In this respect, only beryllium, aluminium and magnesium could be considered. Of these beryllium was too brittle to be used. Aluminium reacted with the fuel at quite low temperatures to form pimples of UAl_2 which burst the cans. Only magnesium was suitable in every respect.

Of the alloys considered, magnesium–manganese alloys were unsatisfactory for the large volume of canning material required, as the manganese additions raised the neutron cross-section unduly. These alloys were of use however for the end-caps and supports. Zinc additions were also undesirable because of the high radioactivity associated with ^{65}Zn, which would have led to troubles in the fuel element cooling ponds. The obvious choice for the canning material was a magnesium–zirconium alloy containing some 0·5% zirconium, which was known to have a fine stable grain size and to be readily workable. The depth of skilled manpower available in the United Kingdom Atomic Energy Authority was such, however, that there were always pressures to develop completely new alloys even when apparently suitable alloys were already available. Consequently an alloy known as AL80 was used, containing some 0·7% aluminium and 0·01% beryllium. AL80 turned out to be an almost ideal canning material which could be extruded or machined to the complex shapes required, had excellent oxidation resistance and good ductility, and generally gave a more than adequate performance to very high levels of fuel burn-up.

The magnesium–zirconium alloy adopted by the French for a similar purpose was in many ways superior. It had an inherently finer grain size, needed less control in manufacture, and was more ductile and less prone to intergranular cavitation than AL80. For an unexpected reason, however, it proved less satisfactory as a can. Plutonium formed by the fuel by neutron capture could diffuse through the magnesium–zirconium alloy, and when outside the can it gave signals to the burst can detection gear which were indistinguishable from the signals from burst cans. Fuel elements thus had to be removed at very low burn-up levels. With AL80, on the other hand, the plutonium reacted with the aluminium to form a very stable intermetallic compound; this prevented plutonium from diffusing to the outside of the can. Thus the choice of the aluminium-bearing alloy proved to be a triumph of metallurgical inspiration.

The main mechanical requirement of magnesium alloy fuel cans is to act as a plastic, deformable envelope to protect and contain the fuel which swells as it is burned-up. High ductility is thus essential. The end caps and end supports, however, are required to support the weight of the fuel elements stacked above them; in the case of these components, therefore, high creep resistance is essential. Magnesium–manganese alloys containing 0·5–1·5% manganese were used as well as magnesium–0·5% zirconium alloys which had been heat treated to improve their creep resistance. The heat treatment, consisting of 96 hours at 600°C in a carbon dioxide environment, is described in Section 5.

(6) MISCELLANEOUS LIGHT-WEIGHT PARTS

The low density characteristics of magnesium alloys are such that they are of interest in the manufacture of many portable pieces of equipment. These include lawnmower decks, scooter fairings, typewriter and copying-machine housings, hand-operated power tools, hand trucks, personal luggage, cine-cameras, projectors and step-ladders. This type of application is most common in the U.S.A., where the standard of living is such that consumers are prepared to pay more for the convenience of low weight.

The low inertia of magnesium components is also an advantage for some applications. These include moving beams used in various types of machinery, particularly in textile production, pulley wheels and centrifuge heads.

11. Conclusions

A survey[2] of the technical and commercial significance of magnesium highlights such factors as the unlimited availability of raw materials and the great advantages which the low density of the metal confers on a wide range of products. The extent of this range depends to a diminishing extent on military applications. It is anticipated that magnesium will become increasingly competitive in price compared with aluminium and the annual growth rate of around 8% in the industry will almost certainly be maintained and probably exceeded. Already, with high volume production, some magnesium components are cheaper than aluminium, and it is probably in the field of large quantity die-casting that the biggest expansion in magnesium markets will occur. In the long term, as ore reserves of many important metals diminish, the unlimited quantities of magnesium available from the sea will ensure a steady growth in the utilisation of the metal.

REFERENCES

1. C. J. P. Ball, *Journal of the Institute of Metals*, vol. 84 (1955–1956), 399.
2. G. Campbell, *Metals and Materials*, vol. 5 (May 1971), 169.
3. American Society for Metals, *Metals Handbook*, vol. 1, 1086.
4. E. F. Emley, *Principles of Magnesium Technology*, Pergamon Press, 1966.

INDEX